João Paulo Greco Ferreira

Influence of plant regulators on sugar sorghum cultivation

João Paulo Greco Ferreira

Influence of plant regulators on sugar sorghum cultivation

Study on the cultivation of sugar sorghum and the application of plant regulators

ScienciaScripts

Imprint

Any brand names and product names mentioned in this book are subject to trademark, brand or patent protection and are trademarks or registered trademarks of their respective holders. The use of brand names, product names, common names, trade names, product descriptions etc. even without a particular marking in this work is in no way to be construed to mean that such names may be regarded as unrestricted in respect of trademark and brand protection legislation and could thus be used by anyone.

Cover image: www.ingimage.com

This book is a translation from the original published under ISBN 978-613-9-72834-3.

Publisher:
Sciencia Scripts
is a trademark of
Dodo Books Indian Ocean Ltd. and OmniScriptum S.R.L publishing group

120 High Road, East Finchley, London, N2 9ED, United Kingdom
Str. Armeneasca 28/1, office 1, Chisinau MD-2012, Republic of Moldova, Europe
Printed at: see last page
ISBN: 978-620-7-89975-3

To my parents Paulo César Ferreira and Maria de Fátima
Greco Ferreira (in memorian) and my siblings Júlia and
Pedro Augusto.

ACKNOWLEDGEMENTS

Firstly, I would like to thank Professor Ronaldo Viana for his friendship, guidance and for believing in my ability, as well as Professor Hildo Sena.

I would like to thank my beloved mother, Maria de Fátima Greco (in memorian), for her example of kindness and dedication. To my father, Paulo César Ferreira, for his trust, support and for being a great inspiration for all my determination and courage.

To my siblings Júlia and Pedro Augusto, for all their affection, trust and friendship.

To my super girlfriend Isabela Camata for her support and trust in me and for believing in my ability.

To my undergraduate friends, especially Eduarda Biagi, Beatriz Leite and Marcos Natalino, I would like to thank you all for your friendship and for sharing moments of learning.

To all the Fatec Araçatuba teachers I had throughout my career as a student.

The Alcoazul plant, for giving me the opportunity to be a trainee at the company and for expanding my knowledge in the area, especially Fabrício Semeão, for his trust and great friendship, and all the other employees who helped me learn.

And last but not least, I thank God, who gave me all of this.

"**Climb the** first step with faith, you don't need to see the whole staircase. Just take the first step" Martin Luther King

3

SUMMARY

Amongst the wide variety of raw materials for ethanol production, sorghum is the most important in terms of productivity when compared to sugar cane. With its high levels of fermentable sugars, rapid reproductive cycle, ease of mechanisation and harvesting that can be brought forward by up to four months compared to sugar cane, this crop stands out from other raw materials. In addition, sorghum can be used to supply raw materials during the off-season at sugar and alcohol mills, supporting national ethanol production. The use of plant regulators in cultivars has the function of altering the physiology and morphology of plants, causing quantitative and qualitative permutations in their formation. They can also restrict plant growth, causing sucrose to accumulate in the stalks. For sugar cane, their application in production allows for greater flexibility in harvest management, as well as ensuring the industrialisation of a better quality product. With this in mind, the use of plant regulators can be seen as an alternative for sugarcane producers and energy production plants, providing an alternative for the production of biofuels. Studies on the use of plant regulators in the cultivation of sugar sorghum are scarce. The aim of this study is therefore to develop a study on the cultivation of sugar sorghum and the application of plant regulators, obtaining a survey through bibliographical works of the main concepts for improving the production of the crop, taking sugar cane as a reference, adapting to the new cultivation procedures, with a view to improving productivity.

Keywords: Sorghum bicolor (L.) Moench, plant regulator, ripeness, ethanol.

SUMMARY

INTRODUCTION

Sorghum [Sorghum bicolor (L.) Moench] is considered to be a grass that varies greatly in its use, with certain cultivars for different products. It is drought tolerant and can adapt to large parts of Brazil's regions. At present, the main focus on sugar sorghum is to obtain sugars from its stalks, which bring great advantages to the bioenergy sector (FILHO et al, 2008).

It originated in Africa, although some evidence suggests that it spread to India. According to archaeologists, its domestication took place around 3000 BC, when the cultivation of other cereals and the practice of domestication were incorporated into Ancient Egypt from Ethiopia. In Brazil, sorghum cultivation has now been incorporated, with commercial consideration beginning in the 1970s (RIBAS, 2003).

According to Oliveira (2012) sorghum has a short growing season, requiring a maximum of 120 days to be planted and harvested. It is a crop that can be sown between November, December or in February, where it would occupy areas that would have been renewed from sugar cane cultivation, which should take place over a period of five years, which currently constitutes cattle grazing areas, mainly in Mato Grosso do Sul, in the northern region of Paraná and in the north-west of São Paulo.

Sorghum has a strategic influence on the distribution of grains and fodder in the country, being able to help in the development of livestock by ensuring the supply of food. Among the wide range of raw materials used to produce ethanol, sorghum is highly regarded in terms of productivity when compared to sugar cane, due to its high levels of fermentable sugars, rapid reproductive cycle, ease of mechanisation and the fact that it can be harvested up to four months earlier than sugar cane. As a result, this crop stands out from other raw materials. As a result, sorghum can be used to supply raw materials during the off-season at sugar and alcohol plants, supporting national ethanol production (COELHO, 2010).

According to Pontin (1995), natural ripening can fail, even in immature varieties. In this context, the role of chemical ripeners is emphasised as an instrument of great importance.

The use of plant regulators in cultivars has the function of altering the physiology and morphology of plants, causing quantitative and qualitative changes in formation. They can also restrict plant growth, causing sucrose to accumulate in the stalks. For sugar cane, its application in production allows for greater flexibility in harvest management, as well as ensuring the industrialisation of a better quality product (CAPUTO; BEAUCLAIR, 2005).

With this in mind, the use of plant regulators can be seen as an alternative for producers and sugar and ethanol plants, providing an alternative for the production of biofuels.

Studies on the use of plant regulators in the cultivation of sugar sorghum are scarce. The aim of this study is therefore to develop a study on the cultivation of sugar sorghum and the application of plant regulators, taking sugar cane as a reference, adapting to the new cultivation procedures, with a view to improving productivity. A survey of the main concepts for improving the production of sugar sorghum was then carried out using bibliographical works.

CHAPTER 1

LITERATURE REVIEW

1.1 History of sugar sorghum in Brazil

In 1975, the Brazilian government decided to use ethyl alcohol as a fuel, hoping that the country would be able to increase its agricultural study area without reducing the areas used for other crops. This proved to be a viable option, as it would reduce the cost of importing oil. This led to the creation of the National Alcohol Programme (Proálcool) by decree no. 79.953, of 14 November 1975, with the aim of encouraging the production of ethyl alcohol, where at that time sugar cane was the main raw material used (RODRIGUES, 2010).

In 1977, the National Maize and Sorghum Research Centre (CNPMS) added some variations of raw materials to its genetic improvement programme, with the aim of producing a purer biofuel. This led to the development of new cultivars of sugar sorghum, adapted to the variations in the Brazilian climate and soil, in order to increase crop production and profits (ANDRADE, 1988).

In Brazil, sugar sorghum can be highlighted as a species of technological command due to the fact that it produces food and energy, restores sugarcane plantations, is highly efficient in agricultural parameters, maximises the use of certain resources such as energy balance, water, soil, low amounts of carbon in agriculture and may be able to produce more biomass during the sugarcane off-season (DURÃES et al, 2012).

Figure 1 shows a sugar sorghum plantation.

Figure 1: Sugar sorghum
Source: COELHO, 2011.

During the first year of the programme, there were a number of factors that limited the improvement of this grass. One of these was the amount of seed available for cultivation. The varieties of fodder sorghum that proved to be of the saccharine type showed low efficiency in the degree of sugars retained in the stalk, and the cultivars were Honey, which stood out for use in the north-east of Brazil, and Start for the south (SCHAFFERT et al, 1978).

Over the last 10 years, sorghum production has grown, especially in the Centre West and Southeast, during the harvest season.[a] Sorghum is often grown later than maize, in the months of September to November, because it is more resistant to the climate, soil varieties, among other things, and can be grown at different times (BAHIA FILHO et al, 2008).

As the renewable energy market appears to be very busy, sorghum cultivation is tending to increase its use, as it can compete with sugar cane in the production of renewable fuels. The United States, India, China and some African countries see sugar sorghum as a favourable option for mastering the production and use of bioethanol (MARCOCCIA, 2007).

The increase in the size of sugarcane sorghum in the regions of Brazil shows great prospects and takes several forms. By expanding the preference for sugarcane sorghum during the sugarcane off-season and/or the reform of sugarcane plantations, with a focus on ethanol production, prematurely proposing the optimum raw material, and maximising the use of the sector's industrial regions, machinery and resources, focusing on competition with sustainability, based on increases in production and decreases in production costs. This is a necessity for the sugar-energy sector and therefore a real chance for the expansion of sugar sorghum in Brazil.

It can be said that the market is very favourable to competitive interests in terms of technology (DURÃES et al, 2012).

According to Conab (2015), the main use of sorghum in Brazil is to produce grain for animal feed. Currently, its production has been increasing both in terms of production and the number of areas planted. The 2014/2015 harvest produced 1992.1 million tonnes of grain on 734.4 million hectares of land, with a yield of 2.71 tonnes per hectare.

The main growing region is the Midwest, which accounts for 61.3% of the national harvest, producing 1.126 million tonnes. This is followed by the Southeast with 22.9 per cent, the Northeast with 11.5 per cent, the South with 2.5 per cent and the North with 1.8 per cent. While the Centre West is the largest producer of grain sorghum, the South and Southeast stand out in terms of fodder sorghum production.

Figure 2 shows the map of sorghum agricultural production in Brazil according to Conab data between the 2014 and 2015 harvests.

Figure 2: Map of sorghum agricultural production in Brazil
Source: Conab, 2015.

According to Conab 2015, on farms where fodder sorghum is planted, there are different traditions among its growers, due to its high productivity and quality. With good management, it can reach an average of 50 tonnes of green mass per hectare, with a range of up to 80 tonnes, as recorded by studies in Goiás. In addition to sorghum's high productivity and quality, it can also be used as a second cut or regrowth, eliminating a high cost for the farmer, and can reach up to 20 tonnes.

The development of sorghum in Brazil has priority in areas that are highly flexible to water shortages, in the Southeast, South and Centre-West regions. And it has the capacity for cultivation in the Northeast, a region that is currently cultivated mainly by small-scale producers using a consortium system. Between the 2013 and 2014 harvests, Bahia was the largest producer in the Northeast region, with 118.9 tonnes per hectare, and the Pernambuco region with a production of 1.0 tonnes per hectare, according to information from the National Supply Company (CONAB, 2015).

As far as Brazil is concerned, it is of great importance to have a space occupied by sorghum, in order to guarantee the supply of grain. In some years, due to negative events, there has been a deficit in the grain and sorghum industry, considering that it is a crop that can be grown in a variety of soil and climate conditions. As a result, some industries in partnership with Embrapa Maize and

Sorghum are carrying out research aimed at acclimatising and developing certain cultivars, particularly forage sorghum and grain sorghum, which are highly productive and tolerant of abiotic and biotic irritants. This work has led to the possibility of obtaining cultivars with associated values that make it possible to improve the performance of this crop (EMBRAPA, 2010).

Currently, areas of the sugarcane agro-industry, including sugarcane farms and the agricultural machinery production industries, have been seeking to qualify sugar sorghum, with a view to new work focused on greater competitiveness and sustainable development in the sugar-energy sector. Sugar sorghum is considered to be an agricultural species with very rudimentary characteristics, which portrays excellent adaptations to the stresses of nature, such as variations in climate and humidity. It is very malleable when it comes to the application of modern elements, such as fertilisers, water and agricultural correctives, at critical points in the flowering and evolution of the crop (COELHO, 2011).

The genetic adaptation of sugar sorghum species and hybrids, technical suggestions on agro-industrial production systems, the shortage of quality seeds, together with certain opportunities for production arrangements, are analytical items for the introduction and expansion of sugar sorghum in the sugar-energy sector (DURÃES et al, 2012).

The mastery of seeds with their adapted genetics is fundamental to making a profit in terms of productivity and appropriate techniques for handling and planting the crop, such as the size of the planting furrows. The constitution of the juice, as well as the production of sucrose and total sugars and the distinction of residues, require extreme care in terms of the use of industrial and technical processes that can lead to an increase in economic results. The basis for progress in agronomic technologies is based on these points of view: good agricultural practices, plant descriptors for energy purposes and both genetic and developmental improvement (COELHO, 2011).

The increase in the cultivation of sugar sorghum in Brazil has great prospects and accepts different models. The model in which sugar sorghum is grown in the sugarcane off-season or in the restructuring of sugarcane plantations aims to add to ethanol production by providing quality raw materials and increasing the utilisation of the industrial park, focusing on competitiveness and sustainable production, aiming to increase production, lower production costs and increase the efficient use of resources and inputs (DURÃES et al, 2012).

1.2 Ethanol production from sugar sorghum.

Alternatives for generating renewable energy and biomass are in great demand these days. The use of sugar sorghum to produce ethanol has been known for some time in Brazil. Since the start of the Pro-Alcohol programme (four decades ago), this means of generating ethanol has been studied and focused on micro-distilleries in particular, and due to the increase in demand for ethanol and bioelectricity, this opportunity has been reactivated (SORDI, 2011).

The use of sugar sorghum in ethanol production is being examined and new hybrids and varieties are being perfected with the aim of increasing ethanol production per hectare. The utilisation of the complete raw material (bagasse, straw and juice) can be considered one of the most important reasons for making ethanol viable. The formation of these phases will delimit the stages in the process of utilising the conversion of biomass into ethanol (MACHADO, 2011).

The raw material represents the start of the industrial process and must undergo quality controls to ensure a higher yield and therefore greater efficiency. In order for ethanol production from sugar sorghum to be homogeneous and controlled, the extraction of the juice is a very important point to focus on. The sorghum bagasse generated after milling can also be used, just like sugar cane bagasse, to cogenerate energy by burning it in boilers and/or to produce second-generation ethanol (GOMES et al, 2011).

Sugarcane is the raw material most used to produce alcohol. However, some Brazilian states have no alcohol production or don't produce enough to meet domestic demand. One example is Rio Grande do Sul, which produces only 2% of the alcohol it consumes. Despite this, there is a clear increase in the number of biofuel-powered vehicles in the country, and with this fleet expected to increase, the demand for alcohol should be pressurised in the following years (EMYDIO, 2010).Figure 3 shows production indicators for sugar sorghum and sugar cane.

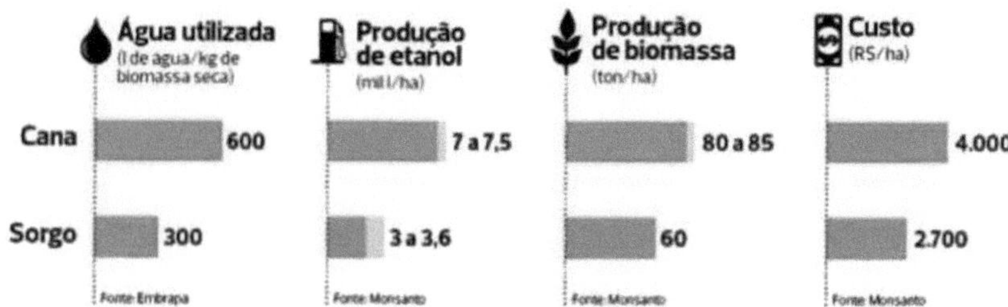

Figure 3: Sugarcane x Sorghum.

Source: GILMAR, 2011

According to Oliveira (2012), the use of sugar sorghum to produce ethanol complements the sugar cane plantation, which is harvested between April and November. With a shortage of ethanol in March and December, the price of ethanol increases in the off-season, distancing consumers who have vehicles fuelled by this fuel. One of the great advantages of sorghum is the juice extracted from the stalks, which is used very well in the industrial process of sugarcane mills where ethanol is produced. All that is needed is a minimal modification to the equipment and a few specific adjustments to the sugar cane harvesting machines so that they can be used for processing and harvesting sorghum.

With its shorter life cycle, the cultivation and management of sorghum is quicker than that of sugar cane, so there is a need for a group of workers dedicated to sorghum (SORDI, 2011).

Table 1 shows the parameters used to compare the two raw materials, sugar sorghum and sugar cane, for ethanol production.

Table 1: Comparison of sugar sorghum and sugar cane parameters for ethanol production.

Parameters	Sugar sorghum	Sugarcane
Crop cycle (months)	4 - 4,5	9 - 18
Water demand (m)^3	9.000	36.000
Stalk yield (tonnes/ha)	35 - 70	75 - 90
Ethanol production (l/ha)	1.400 - 2.800 2.800 - 5.600 (a)	5.600 - 6.500
Geographical coverage (latitude)	52° N to 40° S	30° to 35° S

(a): consideration of two crops/year

Source: EMYDIO, 2010.

It can be seen in Table 1 that sugar cane contains higher numbers in the parameters compared to sorghum, although the crop cycle of sorghum is shorter than that of sugar cane.

According to GOMES et al (2011), many studies show that the essential sugars present in sugar sorghum juice are sucrose, glucose and fructose. Sugar sorghum has a higher content of reducing sugars when compared to sugarcane juice, which is why studies show that there is no great difference in the total amount of soluble sugars. As a result, the same analytical means and parameters that are used for quality control in sugar cane can be used to analyse sugar sorghum juice, namely the gross percentage of PCC sugar, pH, purity of the juice (%), industrial fibre of the material (%), sucrose (Pol of the juice in %), soluble solids (Brix in %). With the production of ethanol using sugar sorghum juice, its primary quality control analyses are already those used routinely in the mills.

According to Durães et al (2012), Embrapa Milho e Sorgo has developed a cultivar that demonstrates excellent adaptation to numerous Brazilian soil and climate conditions, the BR 506 cultivar. According to the researchers, the ethanol yield potential from this cultivar is considered at three technological levels of handling, as shown in Table 2.

Table 2: Ethanol yield of the BR 506 cultivar, based on three technological levels of crop management.

Levels	Green Mass (t/ha)	Green Mass Ethanol (l/t)	Ethanol per area (l/ha)	Cycle (months)	Ethanol yield (litres/ha/month)
Bass	40	50	2.000	4	500

| **Mediu m** | 60 | 60 | 3.600 | 4 | 900 |
| **High** | 80 | 70 | 5.600 | 4 | 1.400 |

Source: DURÃES eta al, 2010.

Table 2 shows that the higher the technological level in a four-month cycle, the higher the ethanol production yield for the BR 506 cultivar.

The quality and specification of the ethanol produced from sorghum, as well as the calorific value of its bagasse when used for burning in boilers, were highly compatible with sugar cane. Its energy balance was also positive. This shows that its fibre and extraction yields can generate enough bagasse for the process. However, it is necessary to assess on a large scale the real amount of electrical energy it will provide (SORDI, 2011).

According to Machado (2011), in Brazil and most of the world, ethanol is obtained through fermentation. In this process, any product with a considerable amount of carbohydrates in its raw material has the potential to produce ethanol. Based on carbohydrates, the raw materials are divided into:

- Sugar: made up of simple sugars, which are carbohydrates with six carbon atoms, such as fructose, glucose and sucrose. The fermentation of these sugars produces ethanol, making it a simple and cheap process;
- Starch: these are complex carbohydrates, such as starch, which can be broken down into glucose by acid hydrolysis or by the action of enzymes in a procedure known as saccharification or malting, although this process is more expensive than the sugar method;
- Cellulose: not an economically viable process, at least at the moment, as it is a very complex means of production and requires physical and chemical pre-treatment.

Sordi (2011) emphasises in his study that by improving certain problems, it is undoubtedly possible to improve them and achieve optimum results. These opportunities for improvement in sorghum for ethanol production are:

- Improve varieties/hybrids that are vigorous to herbicides, borers and tumbling;
- Use of transigia to control the borer;
- Increase agricultural production by improving the number of plants per hectare;
- Varieties/hybrids with larger stalk diameters;
- Minimising harvest losses;
- Varieties/hybrids that limit isoporisation and flowering.

15

CHAPTER 2

SACARINE BLOOD

The following chapter covers a general overview of sugar sorghum, from its treatment for planting to harvesting methods.

2.1 Growing sugar sorghum

With regard to the regions with available capacity for growing this species (sugarcane ratoon regions), the first action to be taken is undoubtedly to obtain an optimum soil sample, with depths of 0 - 20 and 20 - 40 cm, in order to correctly analyse the amount of lime and nutrients to be applied and whether there is a need to correct the soil (MAY, 2011).

According to Coelho (2011), cultural treatments have a direct influence on the entire crop's productivity, so favourable circumstances for the production of the various sorghum species (sugar, grain, fodder and broom) make up several elements for excellent development. Such as:

a. planting period;
b. climate;
c. fertilisation;
d. relationships between soil characteristics;
e. crop rotation and succession techniques;
f. phytosanitary treatments (dispersal of pests that damage the crop).

Figure 4 shows some of the cultivation methods for producing sugar sorghum.

Lavoura de sorgo | Colheita mecanizada | Plantio tradicional

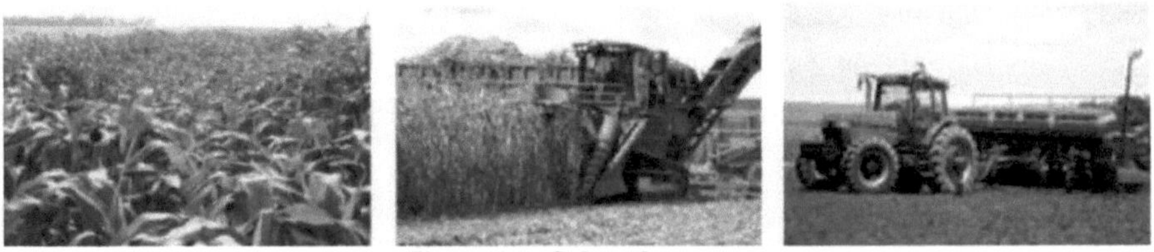

Figura 4: Cultural treatments

Source: MAY, 2014.

Figure 4 shows three cultivation treatments which, when used correctly, have a major influence on crop yields.

4.2 Sowing periods

Most of the time, it is advisable to sow sugarcane sorghum in sugarcane-growing regions, such as the Southeast and Centre-West, between November and December. Harvesting takes place in March and April, when sugar cane does not show a high soluble solids content (brix), preventing it from being cut (MAY, 2011).

The planting period is of great importance as the cultivars used have an affective photoperiod, which includes variations (short days) and brings down the crop cycle, leading to negative factors in terms of yield (BORGONOVI, 1982).

According to Mantovani (2000), a considerable point when planting sorghum is to regulate planting density, where excellent regulation will favour maximum crop efficiency, which can vary greatly depending on the cultivar and the idleness of nutrients and water. A reduction in planting density can occur in regions where there is a large amount of straw (poorly distributed straw) and uneven micro relief, which can delay the initial development of the plant. In order to achieve better development of sorghum crops with higher yields, sorghum needs optimum conditions just like any other plant species, such as: pest and weed control, fertilisers, adequate soil preparation, among many other points (MAY, 2011).

4.3 Fertilisation

According to Coelho (2011), the implementation of good conditions for producing sorghum requires knowledge and the management of various elements. The productive potential of sorghum is determined by the characteristics of the soil, the climate, the crop rotation system, the planting period, fertilisation and plant health treatments. Therefore, a rational fertilisation programme must take these considerations into account:

- Diagnosis of soil fertility;
- Patterns of nutrient absorption and accumulation, with the main focus on nitrogen, potassium and phosphorus;
- The plant's nutritional requirements, according to the purpose for which it is grown;
- Manipulation of fertiliser;
- Source of nutrients.

The application of nitrogen to sorghum crops varies according to the level of organic matter in the soil, and can reach an application of 20kg per hectare and the rest in top dressing, from the moment the plants have five to seven leaves (around 30 to 35 days after emergence), before the flower buds begin to differentiate. The climate has a direct effect on this top dressing, which can be partially or totally cancelled out (RODRIGUES et al, 2010).Each type of soil has different characteristics, which differs in the factor in which the nutrients should be received. It is therefore necessary to quantify, through laboratory analyses, the potential of the soil to supply nutrients and the nutritional status of the plants. This provides a basis for the correct and efficient use of fertilisers (COELHO, 2011).

Since sorghum is in its introductory period, the types of fertiliser recommended are based on those used on maize, grain sorghum and forage sorghum, so the effect has been more or less permanent for nitrogen and phosphorus, but less so for potassium. These demands are greater when used on sugar sorghum (BORGONOVI, 1982).

4.4 Zoning

Obtaining a satisfactory zoning has to be done in regions that are suitable for ethanol production, in order to provide subsidies for the future definition of territorial study tactics. For both sorghum and sugar cane, agro-ecological zoning has inequalities and restrictions from one crop to the next, requiring the identification of the region, its climatic resources and a study of the soil, so that the land is earmarked for agricultural use and does not demonstrate environmental restrictions that could be detrimental to the crop to be planted (LANDAU; SCHAFFERT, 2011).

For Manzatto (2011), the zoning instructions for sugar sorghum can be the same as those for the National Agroecological Zoning of Sugarcane (ZAE Cana). It is possible to separate four considerable perspectives:

1. Qualify, indicate and adopt knowledge of the potential of soils for generating sugar sorghum;
2. To provide technical assistance in sowing sugar sorghum with a view to increasing ethanol production;
3. Knowledge of the soil's agricultural potential;
4. To identify regions suitable for increasing the cultivation of sugar sorghum varieties.

Zoning provides farmers with guidance on the ideal periods for planting any crop in various areas of the country. This gives farmers a certain degree of certainty when planting, reducing the risk of crop losses due to climatic variations, especially during the most critical periods of crop development (LANDAU; GUIMARÂES, 2010).

Figure 5 shows a map highlighting the areas that could potentially be planted with sugar sorghum during the sugarcane off-season in Brazil.

Figura 5:   Potential areas for planting sugar sorghum during the sugarcane off-season in Brazil.

Source: LANDAU; GUIMARÂES, 2010.

2.5 Weeds

Any plant that is in a certain environment and is not wanted can be considered a weed or invasive plant. They are plants that have no genetic improvement and show great resistance to adverse soil and/or climate conditions. They are also resistant to pests and diseases and produce a large quantity of seeds, making them easy to disperse (OLIVEIRA, 2011).

It is advisable to grow sorghum in weed-free soil so that the plant is not smothered. If these weeds are not removed during the emergence period, sorghum yields can be reduced by up to 25 per cent. Weed control usually involves the application of herbicides or mechanical means (BORGONOVI, 1982).

One of the major problems in growing sugar sorghum is controlling weeds, which affect the crop because they compete for the same goods: light, water and nutrients, especially nitrogen (COELHO et al, 2002).

2. 6Row spacing   and planting densities

Photosynthesis allows the plant to produce between 90 and 95 per cent of its dry mass and is the method by which the plant has enough metabolic energy for its evolution. Throughout the sorghum cycle, from germination to flowering, the leaves are the main photosynthetic organs, so the size of the leaf area directly affects the amount of photosynthesis. The use of smaller row spacings is advantageous because it results in better yields, covering the soil more quickly and benefiting greatly from the omission of invasive or weed plants, reducing infestation (RODRIGUES, 2010).Figure 6 shows the row spacing that is most advisable for planting sugar sorghum.

Figura 6:   Spacing used for planting sugar sorghum

Source: DEDINI, 2010.

Reducing planting spacing and increasing plant density makes it easier to intercept solar radiation because it increases the leaf index, which also increases water utilisation, dry mass production and nutrients (MOLIN, 2000).

According to Sordi (2011), most herbicide molecules are very sensitive, with atrazine, a herbicide used to control the pre- and post-emergence of plants infested with other crops, generally sugar cane, maize and sorghum. It obstructs weed control, because in mechanised harvesting, using sugar cane machinery, the spacing used is different, leading to larger inter-row areas. In order to better define the vegetation population in the area, studies have been carried out to provide better agricultural productivity and weed management, generating less loss when harvesting is carried out.

2.7 Phytosanitary product applications

Research is being carried out in order to interpret the problems and find ways of reducing them. Of the ways found, the most frequently used to protect sorghum, sugar cane and other crops against pest and weed attacks is the use of phytosanitary products, applied both to the plant and directly to the soil, with spraying and handling being the most widely used (COSTA et al, 2009).

The use of phytosanitary products is considered one of the most important phases when it comes to rural production (CARVALHO, 2003).

According to Costa (2009), the successful use of phytosanitary products in agriculture depends fundamentally on the application of products with optimum efficiency and perfected application technology, as well as the timing and influence of uncontrolled organic, meteorological and biological factors.

Agricultural pesticides have evolved over time. This has led to the emergence of products that are considered less toxic to both the environment and humans. However, their application must be carried out with suitable equipment and competent workers, without any undesirable repercussions (BOLLER, 2004).

According to Costa (2009), there are a number of points in the current methods of using pesticides that are considered essential if we are to be successful in protecting crops, cancelling out the manifestation of pests and also the competitiveness of other crops:

-Climatic conditions;

-Drop sizing calculation;

-Operating criteria;

-Equipment   ;

-Timely period.

2.8 Harvesting

The accumulation of sugars in larger quantities begins after flowering. It continues until the ripening stage is completed. During this period, the ART (Total Reducing Sugars) show high rates in their juice and in the percentage of extractable juice, these parameters are intended to determine the correct point of harvest, and can be influenced by environmental conditions and the type of cultivar. These two points can determine the correct point of harvest, which can be verified through the values of total reducing sugars, brix and the percentage of juice in the ripening curve of each cultivar (SCHAFFERT et al, 1980).

During the 2010-2011 and 2011-2012 harvests, sugar sorghum was harvested using the same system as sugar cane, with losses of less than 1%. On some occasions, the machinery had to be adjusted so that the whole plant could be taken to the mill, but the amount of impurities in the processed sorghum was high, causing milling problems for some mills (MAY, 2013).

Figure 7 shows sugar sorghum being harvested using sugar cane harvesting machinery.

Figure 7. Mechanised harvesting of sugar sorghum.
Source: OLIVEIRA, 2012.

Figure 7 shows that it is possible to harvest sugar sorghum using sugar cane harvesters.

CHAPTER 3

PLANT HORMONE

This chapter will study the main plant hormones, their action and influence on the plant.

Plant hormones are also considered to be one of the major agents capable of organising the genetic potential of plants, acting to correct essential genes. A plant hormone or phytohormone is defined as an organic element, agile in very small concentrations, formed in certain regions of plants and transported to other areas, causing biochemical, physiological or morphological responses (RODRIGUES; LEITE 2004).

Hormones are transported by the two sap-conducting vessels that exist in plants (Xylem and Phloem) and can act directly in the area where they are produced. The main hormones found in plants are Auxins, Cytokinins, Gliberelins, Ethylene and Abscisic Acid, which can fulfil various functions such as flowering, ripening fruit, stimulating plant growth and forming all its tissues. They can also help the plant adapt to conditions of water scarcity, water stress and induce germination and/or break dormancy (MAGALHÃES; VIERA 2008).

According to Rodrigues & Leite (2004) the action of plant hormones is normally quite small, in the range of 1 pM to 1 mM, and their greater efficiency is due to their mobility throughout the organism, their potential for amplifying signals and their ability to achieve more complex regulatory action through interactions with many other biochemical and physiological processes.

3.1 Auxin

Auxins have many effects on plant growth, including stimulating the elongation of stems and cleoptiles, fruit development and more. The effects of auxins are influenced by many factors, including: the type of auxin, the concentration, the stage of organ development, the presence of other plant hormones and the condition of the tissue (MAGALHÃES; VIEIRA 2008).

According to the degree programme in exact sciences at the São Carlos Institute of Physics, any substance that can stimulate phototropic curvature, stems and various plant structures can be considered an auxin. In plants, the primary auxin is IAA (indoleacetic acid). IAA is generated in the apical meristem of buds, new leaves and seeds. It is not transported by the xylem and/or phloem, but through the plant between the parenchyma cells, at a rate that is too high to be done by diffusion. The movement of auxin is called apolar transport because it is always unidirectional, from the apex of the meristem to its base. This transport requires energy, and gravity does not influence it.Auxins are considered to be special substances in plant cultivation. They favour changes to the plant's cell wall throughout the process of cell division, where they increase the extensibility of the entire cell. They are capable of stimulating many physiological responses when the focus is on inducing roots, axillary or apical buds, leaves, stems and embryos (SILVA et al, 2013).Figure 8 shows natural auxin and synthetic auxins.

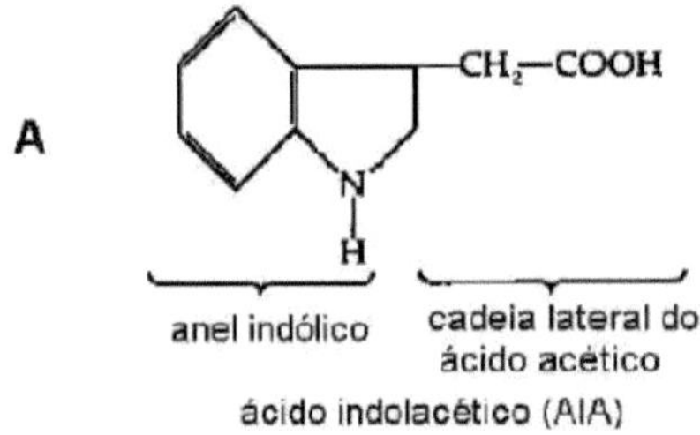

Figure 8: A) natural auxin; B) and C) synthetic auxins.

Source: Introduction to plant biology.

In biochemical terms, auxins act in two periods. At the beginning there is a slight growth stimulus, which appears to be intervened by the efflux of protons and probably through modifications in the calcium of the cytosol, with the action of auxin on already existing cellular components. In the second phase, auxin probably acts by activating and/or correcting specific genes (RODRIGUES; LEITE, 2004).

3.2 Cytokinin

Cytokinins are hormones which, when of natural origin, are found as free molecules, i.e. they do not bind to any macromolecule, and also as remodelled bases in suitable t-RNA molecules. Cytokinins are synthesised in roots, growing embryos and tissues (RODRIGUES; LEITE, 2004).

Like any plant hormone, its action takes place in three phases: understanding of the signal; transduction of the understood signal; and the primary points of hormonal activity. In the first phase, the understanding of the signal is realised by the hormone joining with a specific receptor. Receptors are proteins found on the cell membrane or even in the cytoplasm, where they bind to chemical messengers with a specific and reversible format. Following binding, this receptor may undergo changes, reaching an activated phase in which it triggers major intracellular chemical events. As a result, the signal received must intervene in cellular mechanisms, such as expansion, differentiation or division, which are fundamental factors that contribute to modifying the plant (COSTA, 2006).

According to Rodrigues; Leite (2004) cytokinins have the following physiological effects:

- Incitement of cell growth in certain organs and tissues;
- Co-operation in the maturation of chloroplasts;
- Participation in the adjustment of the plant cell cycle;
- Encouraging the mobilisation of nutrients and delaying ageing;
- Participation in the regulation of morphogenesis in plant tissue culture.

As well as increasing the rate of protein synthesis, cytokinins also modify the spectrum of proteins generated from plant tissues. Not only are there free cytokinins in the cells, there are also cytokinins bound to t-RNAs, which are involved in a critical situation adjacent to the anticodon and can intervene in the bond between the t-RNA and the m-RNA (RODRIGUES; LEITE, 2004).

According to Costa (2006), a favourable oscillation between auxin and cytokinin induces cytokinins to form stem buds. Under normal operations, the concentration of cytokinin substantially induces bud formation, which is not necessarily higher than that of auxin. There is evidence that such a differential effect between auxins and cytokinins when inducing stems and roots is relevant to the complete evolution of the plant.

Figure 9 shows the hormonal balance between cytokinin and auxin, and that a high intensity of cytokinin favours root growth, while auxin in high proportions induces root differentiation in the parenchymatous tissues of the pith (COSTA, 2006).

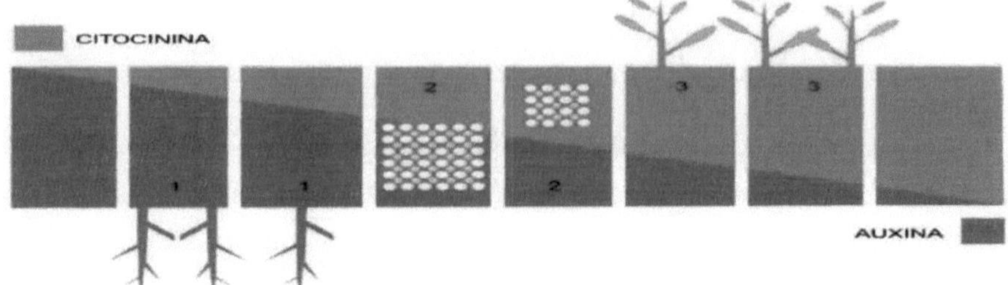

Figura 9:   Interaction between Auxin and Cytokinin.

Source: KERBAUY, 2004.

3.3 Gibberellin

The plant hormone gibberellin is a hormone that is localised in certain areas of plants, such as their roots, young foliage, seeds during the germination period and fruit. Gibberellin acts on the development of the stem and leaves of plants, regulating their height. It also acts on fruit development, flowering and slowing down the ageing of plants. This can also occur through another hormone, auxin. Gibberellin moves in the opposite way to auxin, although it is produced in the same place, as it is apolar and is transported through the xylem and phloem, from the top of the plant to the base (LAVAGNINI et al, 2014).

Gibberellins act throughout the plant cycle, intervening in seed germination, stem elongation, flowering, anther growth and seed and pericarp development. They intervene in environmental stimuli, which is why the adjustment of the gibberellin biosynthesis pathway is of fundamental importance for plant development and its adaptation to the environment. From the moment that gibberellic acid began to be commercialised, it was used on a large number of plants, and at times, great results were obtained, leading one to think that gibberellins could cause a great increase in plant productivity (RODRIGUES; LEITE, 2004).

Figure 10 shows the chemical structure of gibberellic acid (GA3).

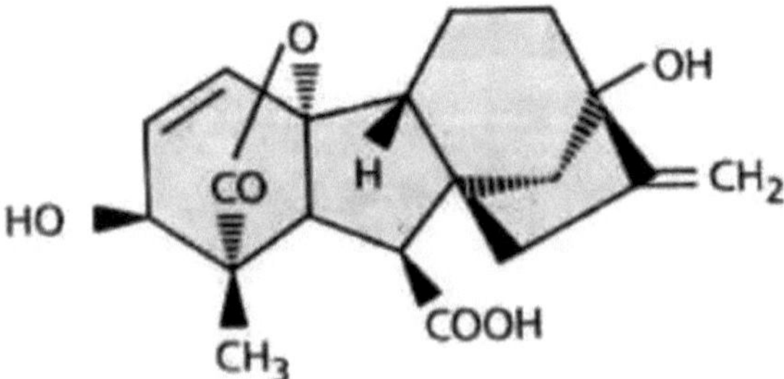

Figure 10: Chemical structure of gibberellic acid.
Source: COSTA, n.d.

When applied to sugarcane, gibberellins can increase the size of the internodes, resulting in an increase in sucrose production. There are results reporting that the application of GA induces an increase of 20 tonnes per hectare in sugarcane stalk production and 2 tonnes per hectare in sucrose production. Growth retardants or inhibitors of GA biosynthesis have been used in agriculture for genetic improvement purposes and are used to reduce plant lodging, tree development, environmental stresses (frost) and to induce flowering (LINDOMAR; HAVANE, n.d.).

According to Rodrigues & Leite (2004) the main effects of gibberellins are as follows:

- Increased flower stalk;
- Altered youthfulness;
- Stem elongation of dwarf plants;
- Induces masculinity in flowers;
- Induces seed germination;
- Fruit growth.

Gibberellin acts on every part of plant development, from breaking dormancy in seed germination, leaf and stem development so that the plant develops fully, to the formation of reproductive organs and the generation of fruit. In certain crops, it has the effect of delaying fruit ripening (LAVAGNINI et al, 2014). In sugar cane, gibberellins favour an increase in production, leading to greater internode elongation, without affecting sugar concentration (RODRIGUES; LEITE, 2004).

3.4 Ethylene

Ethylene is an open-chain hydrocarbon with a molecular weight of 28 Da and is lighter than air. It is found in high concentrations in fruit during the ripening period, cigarette smoke and incomplete combustion vehicles. Some flowers are exposed to small concentrations of ethylene when they die. Its structure is $CH_2 = CH_2$ (RODRIGUES; LEITE 2014).

Figure 11 shows the structure of ethylene.

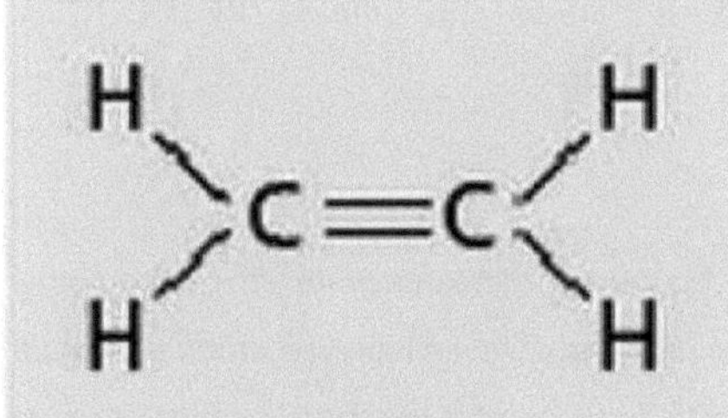

Figure 11: Structure of ethylene.

Source: QUEVEDO et al, 2011.

Ethylene can be produced anywhere in higher plants, but its production depends on the type of tissue and the stage of development. Concentration increases during periods of flower senescence, leaf fall and ripening. It can also be induced by environmental conditions, such as water stress. Plant tissues that have been injured or mechanically disturbed have their ethylene production increased for a period of 30 minutes, returning to normal after some time. Ethylene is easily released from the tissues and is released as a gas through the intracellular spaces (QUEVEDO et al, 2011).

Ethylene is synthesised from methionine and its metabolic intermediates are S-adenosyl methionine and aminocyclopropane carboxylic acid (ACC). Its biosynthesis is influenced by many factors, including stage of development, environmental conditions, other hormones and physical and/or chemical damage. When a plant cell is damaged, ethylene production increases (RODRIGUES; LEITE, 2004). According to Rodrigues & Leite (2004), ethylene has the following main physiological effects:

- Direction of fruit ripening;
- Root induction;
- Control of falling fruit, leaves, flowers and other organs;
- Flowering control;
- Senescence of leaves and flowers;
- Induction of epinasty.

Ethylene is transported independently of vascular tissues, unlike other plant hormones. A gas diffuses easily within tissues, and through intracellular spaces it can be lost to the environment. Water and solutes in the cytoplasm make it difficult for ethylene to move, as its diffusion coefficient is ten thousand times lower than that of air, making it easier to diffuse through the skin of fruit that has sera (QUEVEDO et al 2011).Ethylene is considered one of the most widely used plant hormones in agriculture, as it has many effects on physiological processes. It is difficult to apply in the field because the gas diffuses quickly, which is why the most widely used compound is ethephon, under the trade name Ethrel. It is applied by spraying plants with an aqueous solution (RODRIGUES; LEITE, 2004).

3.5 Abscisic acid

It is popularly known as the plant stress hormone because it regulates plant growth, stomatal aperture, protein synthesis and other biochemical aspects, especially when plants are exposed to water deficiency conditions. It motivates responses in plants to help protect them against stresses (RODRIGUES; LEITE, 2004).

Abscisic acid (ABA) is a natural hormone that is involved in the regulation of gene physiognomy in important plant physiological factors. The hormone ABA regulates facts of great importance in plant development, including the synthesis of accumulation proteins and lipids in seeds, favours dormancy and inhibits the transition from embryogenic growth to germination and vegetative to reproductive growth. Despite this, ABA is involved in the gene regulation of certain aspects of environmental stress, such as water, salt and cold stress (SILVA et al, 2005).

According to experiments, it has been found that ABA affects protein synthesis, which occurs under adverse conditions such as an increase in temperature. In this case, the use of ABA has resulted in the control of transcription, which leads to the production of new proteins; abscisic acid causes a decrease in mRNA levels. It has also been shown that ABA acts on the accumulation of reserve globulin in maize seeds. Its influence consists of initiating the production and suppressing the degradation of this protein, as well as inhibiting early seed germination (RODRIGUES; LEITE, 2004).

Although it is known that ABA is involved in signalling both abiotic and biotic stresses in plants, its interaction with the signalling pathway and the molecular mechanism in which it is involved remain vague (DOMINGUES, 2013).

According to Rodrigues & Leite (2004), abscisic acid has the following main physiological effects:

- Water absorption control;
- Control of stomata closure;
- Growth control;
- Control of seed dormancy and germination;
- Control over the flowering of certain species;
- Control of senescence;
- Control of bud dormancy.

CHAPTER 4

MATURATION

4.1 Sugar sorghum maturation

According to Teixeira et al (1997), unlike sugar cane, sugar sorghum begins to accumulate sugars in the stalks during its reproductive period, where it is mainly concentrated during the physiological maturation of the grains.

Sorghum is considered a short-day vegetable with high rates of photosynthesis. Most sorghum genes require temperatures above 21°C for optimum plant development. During the first part of the crop's growth, which is from germination (EC1) to panicle initiation, it is extremely important that germination, emergence and seedling establishment take place in the shortest possible time, bearing in mind that the plant is reduced in size, has slow initial growth and inefficient control of invasive plants during this period, reducing grain yields (MAGALHÃES et al, 2000).

The second period (EC2) is the phase in which the panicle begins and lasts until flowering. If these various growth processes are affected, they can jeopardise the yield from the development of the root system, leaves and the accumulation of dry matter. In the last growth period (EC3), which is from flowering to physiological maturity, the events considered to have the greatest impact are related to grain filling. During these three growth periods, photosynthesis, cell division and cell growth must be organised with a view to achieving better crop yields (MAGALHÃES et al, 2000).

It is extremely important to understand when sorghum plants pass from the vegetative to the reproductive stage in order to understand the varieties in their production. After the juvenile stage is over, when the plant is insensitive to photoperiod, the inductive stage begins, when it is fragile to this climatic cause. In this period, during inductive photoperiods, the sorghum crop is exposed to physiological changes in its apical meristem, identified by the initiation of the floral meristem (DINIZ, 2010).

According to Magalhães et al. (2000), the height of sorghum is extremely important for its classification and can vary from 40 cm to 4 m. From the stem to the end of the panicle there can be a variation depending on the number of internodes and the distance between the second peduncle and the panicle. The number of nodes is determined by maturation genes and their reaction to temperature and photoperiod. The spacing of the internodes varies according to combinations of four or more genetic actions and the environment. The flowering of sorghum corresponds to its fertilisation, pollination, development and maturation of the grain. This is the phase in which the plant cannot suffer any biotic or abiotic stress, as its floral differentiation is affected by photoperiod and, above all, temperature. Normally, the flower bud forms 15 to 30 cm above ground level (MAGLHÃES et al, 2000).

4.2 Factors affecting the ripening of sugar sorghum

When we compare sorghum to other crops, we see that sorghum needs a small amount of water in order to develop, so that flowering is its most critical period in terms of lack of water. The sorghum crop is well suited to growing in regions where rainfall is unevenly distributed, so it is able to thrive in many environments (DINIZ, 2010).

Its physiological maturity is not uniform. There is a close relationship between the loss of moisture and the increase in dry matter in the seed, for every three fractions of the panicle in BRS 501 sorghum, during the ripening process. The greatest germination and dry matter production occurs between 47 days after seed maturation, with the apical portion occurring 54 days after maturation. Seed germination begins between 6 and 10 days after anthesis (SILVA, 2010).

The need for favourable situations for the growth of sorghum crops comes up against several factors, such as the sowing time, which can influence the crop cycle, because delaying it can lead to the anticipation of physiological stages, devaluing the product with a drop in production (MACHADO et al, 1987).

Because sorghum is of tropical origin, it is sensitive to low night-time temperatures. Its optimum growth temperature is between 33 - 34°C, when it is higher than 38°C or lower than 16°C, its productivity decreases. Very low temperatures can cause a decrease in leaf area, size, tillering, dry matter accumulation and a delay in the flowering period. All of this is due to a reduction in chlorophyll synthesis, especially in the leaves that first appear on the young plant, resulting in a decrease in photosynthesis (DINIZ, 2010).

In terms of its need for light, temperature and water, sorghum is quite different from other crops in that it needs a low amount of water. Its nutritional requirements vary directly with its production potential. It is expected that the plant's nutritional requirements will increase as productivity rates increase, and for sorghum, the greatest need is for potassium and nitrogen, followed by calcium, magnesium and phosphorus. Because sorghum crops, which have higher yields, export and extract large amounts of nutrients, they need fertilisers with different doses. Official fertiliser data for sorghum cultivation in Brazil states that the recommended doses should be segmented according to the programmed productivity. This concept is not extremely important for magnesium and calcium, because if the acidity of the soil is correctly corrected, this is enough for sorghum crops with high yields (MACHADO, 2011).

Looking at the physiological point of sorghum, the yield of its juice can be defined from the balance between gains from photosynthesis and losses from respiration. Photosynthesis depends on the amount of light that is intercepted. Plant arrangement can have an influence on leaf area index, leaf angle and leaf distribution. Because of these characteristics, the organisation of sorghum plants in the field is extremely important for converting solar radiation into active photosynthesis for the production of its juice (ARGENTA et al, 2001).

According to Machado (2011), sorghum producers should be very careful when it comes to the occurrence of pests and diseases in their crop. From planting to harvest, there are a large number of species of insects and pathogens that can be associated with the crop. However, only a few are significant and there are only a few that cause economic harm. It is therefore very important to always keep an eye on the crop in order to identify the species that can be harmful and adopt methods to control them, because pests in the crop are often not related to the level of economic damage, and, if such a minimum level of handling is possible, this varies with the population. In order to facilitate control, constant monitoring of the crop cycle is necessary, and seeking knowledge of the history of the area to be planted is extremely important for efficient pest and disease control.

According to Diniz (2010), the concept of a weed is given to any plant that occurs in an unwanted location. These plants interfere with the crop's ability to grow in a healthy and lively manner. It can be said that depending on how long the weeds are in contact with the crop, the greater the damage they will cause. This may depend on the crop's cycle and the period in which it was planted.

The Critical Period for Prevention of Interference (CPPI) is the period in which the crop needs to be free of infestations of invaders that threaten to compromise its productivity. To determine this period, the phenological stages of the genotypes or their time periods must be taken into account. With regard to sorghum, its phenological stage is usually adopted because the number of days after emergence varies according to a large number of biotic and abiotic factors (LARCHER, 2000).

Like all crops, sorghum needs cultural treatments, and weed control is one of them. Weeds are considered invasive plants because they compete with sorghum, for example, by competing for water, light and nutrients in the soil, and can cause losses in productivity of between 35 and 70 per cent. The competition between invasive plants and the crop of importance can vary depending on the species. In relation to sorghum, it can be considered that its cycle is shorter compared to other crops, which can cause damage to its development if it is not controlled at the right time, which allows weeds to appear and, together with this, a drop in crop yield (DINIZ, 2010).

4.3 The use of plant regulators in sugar sorghum

Plant regulators are chemical substances, either natural or synthetic, that can be used directly on plants, which can alter their vital and/or structural processes by changing their hormonal status. The use of these products has shown potential in the maturation and productivity of certain crops, but their use is not routine. They are small molecules that have the functionality of specific chemical signals in plant cells, and can manage their development and growth (LAMAS, 2001).

Maturators, characterised as plant regulators, are considered to be chemical compounds that can cause morphological and physiological changes in plants, bringing about both quantitative and qualitative changes. They can also inhibit or slow down plant development, increase sucrose content and cause early ripening (ALMEIDA et al, 2003).

Plant regulators are maturators whose action is to inhibit growth, i.e. they prevent the stalks from elongating, without affecting photosynthesis and favouring the agglomeration of sugars in their tissues. They are used on their own, in combination with each other or mixed with different natural biochemical compounds, such as amino acids, vitamins and micronutrients. Their effects vary according to the species, growth stage, concentration and many environmental factors (CASTRO; VIEIRA, 2001).

The use of plant regulators in sugar cane has become an increasingly normal practice within the sugar-energy sector, with the aim of anticipating and maintaining natural ripeness, providing good quality raw material in advance for the industry, while also helping with crop control (GARCIA, 2014).

According to Garcia (2014), the main function of chemical ripeners, when applied to sugar cane, is to induce the plant to transform its reducing sugars more intensely into sucrose, storing them in its stalk. Its use is extremely important for planning the harvest, as it provides the industry with a higher quality raw material. However, its use depends on a range of factors, including climate, technical and economic factors, as well as each cultivar's reaction to its application.

According to the article written by Leite; Crusciol (2008) in which a study was carried out on the application of three types of growth inhibitors to sugar cane crops: sulfometuron methyl, glyphosphate, and organic carboxylic radical compounds + glyphosphate. As a result, all the ripeners slowed growth in terms of plant height without affecting the diameter and number of stalks, and the ripeners influenced the occurrence of flowering and shrivelling. The ripeners influenced the induction of an increase in the total reducing sugar content, resulting in an improvement in the quality of the raw material for the industry.

CHAPTER 5

FINAL CONSIDERATIONS

The main objective of this study was to clarify doubts about the cultivation of sugar sorghum, show the hormones present in the plant and clarify the usefulness of plant regulators, with a view to improving quality and productivity. Although there is a lack of studies on the application of plant regulators to the sugar sorghum crop, research has been carried out based on the sugar cane crop, due to its similarities.

Plant regulators can be defined as synthetic substances that are very similar to plant hormones and can be applied directly to plants, causing changes, improving crop yields and bringing quantitative and qualitative results to the end consumer, which in this case is the sugar-energy industry.

The use of plant regulators has become a viable method for exploiting the production potential of crops. It actively regulates the physiological processes of plants, which provides an economic return.

REFERENCES

ALMEIDA, J.C.V.; SANOMYA, R.; LEITE, C.F.; CASSINELLI, N.F. Agronomic efficiency of sulfometuron-methyl as a ripener in sugarcane cultivation. **STAB Magazine**, v.21, p. 36-37, 2003.

ANDRADE, R. V.; OLIVEIRA, A. C. **Physiological maturation of the stalk and seed of sugar sorghum,** Revista Brasileira de Sementes, vol. 10, n. 3, p. 19-31, 1988.

ARGENTA, G.; SILVA, P. R. F.; SANGOI, L. **Plant arrangement in maize: a state-of-the-art analysis.** Ciência Rural, Santa Maria, v.31, n.6, p. 1075-1084, 2001.

BAHIA FILHO, A. F. C.; GARCIA, J.C.; PERENTONI, S.N.; SANTANA, D.P.; CRUZ, J. C.; SCHAFFERT, R. E. Boosting maize and sorghum production and productivity in Brazil. In: ALBUQUERQUE, A. C. S.; SILVA, A. G. **Tropical agriculture: four decades of technological, institutional and political innovations**. Brasília, DF: Embrapa Informações Tecnológicas, 2008. chap. 2, p. 125.

BOLLER, W. Technical parameters for tip selection. In: BORGES, L. D. **Tecnologia de aplicação de defensivos agrícolas.** Passo Fundo: Plantio Direto Eventos, 2004.

BORGONOVI, R. A.; GIACOMINI S. F.; SCHAFFERT, H. E. **Cultivares,** p. 38, 1882. Available at   : <http://ainfo.cnptia.embrapa.br/digital/bitstream/item/45831/1/Cultivares.pdf>. Accessed on: 11

August 2015.

CAPUTO, M. M.; BEAUCLAIR, E. G. F. The use of chemical ripeners in sugar cane. **APTARegional,**   08 Nov. 2005. Available at: <http://www.agencia.cnptia.embrapa.br/Repositorio/maturadores_000ftvubmkf02wyiv80otz6 x9i722ji0.pdf> Accessed on: 29 Aug. 2015.

OAK, W. P. A. **Performance of a DGPS flow controller for spraying machines.** Thesis (Doctorate) - Universidade Estadual Paulista, Faculdade de Ciências Agronômicas, 2003.

CASTRO, P. R. C.; VIEIRA, E. L. **Application of plant regulators in tropical agriculture.** Guaíba: Agropecuária, 2001, p.132.

COELHO, A. M. **Sorghum cultivation.** Embrapa milho e sorgo. Sep. 2010. Available at: <http://www.cnpms.embrapa.br/publicacoes/sorgo_6_ed/adubacao.htm>. Accessed on: 19 March 2015.

COELHO, A. M. Sugar sorghum: Soil fertility, nutritional requirements and fertilisation of sugar sorghum. **Agroenergia em Revista,** n. 3, p. 18-19. Aug. 2011.

COELHO, A. M. Sugar Sorghum: Agronomic and Industrial Technology for Food and Energy. **Agroenergia em Revista,** v. 2, n. 3, p. 6-7, Aug. 2011.

COELHO, A. M.; WAQU IL, J. M.; KARAM, D.; CASELA, C. R.; RIBAS, P. M. **Be the doctor of your sorghum.** Piracicaba: Potafós, 2002.

COSTA, D. I. **Efficiency and quality of ground and aerial fungicide applications in the control of foliar diseases and grain yield in soya and maize.** 2009. 126f. Thesis (Doctorate in Agronomy) - Postgraduate Programme in Agronomy. University of Passo Fundo. Passo Fundo, 2009.

COSTA, R. C. L. **Cytokinins.** Federal Rural University of Amazonia. Belém, PA. 2006.

COSTA, R. C. L. **Gibberellins.** S.d. Available at   :< http://www.portalsaofrancisco.com.br/alfa/giberelinas/giberelinas.php>. Accessed on: 29 August 2015.

CONAB. National Supply Company. **Brazilian crop monitoring: grains, seventh survey,** April 2015. Brasília: Conab, 2011. Available at: <http://www.conab.gov.br/> Accessed on: 24 June 2015.

DINIZ, G. M. M. Production of *Sorghum (Sorghum bicolor* L. Moench) General Aspects. 22f.

2011. (Master's Degree in Plant Genetic Improvement) Postgraduate Programme in Agronomy. Federal Rural University of Pernambuco. Recife, PE.

DOMINGUES, B. A. **Salicylic, abscisic and jasmonic acids in grapevines submitted or not to the application of TPC *(Thermal Pest Control)* technology.** 76f. 2013. Dissertation presented to obtain the title of Master of Science. **Luiz de Queiroz College of** Agriculture. **Piracicaba, SP.**

DURÃES, F. O. M.; MAY, A.; PARRELLA,R. A. C. **Embrapa Maize and Sorghum.** Sugar Sorghum Agroindustrial System in Brazil and Public-Private Participation: Opportunities,

Perspectives and Challenges. In Documentos. 138, 2012, Sete Lagoas-MG, 2012. Proceedings... Sete Lagoas, EMBRAPA, 2012.

EMYDIO, B. M. **Ethanol production from sugar sorghum.** 2010. Available at: <http://www.infobibos.com/Artigos/2010_4/sorgo/index.htm>. Accessed on: 17 July 2015.

FILHO, J. P. R. do A.; FILHO, D. F.; FARINELLI, R. and BARBOSA, J. C.. **Spacing, population density and nitrogen fertilisation in maize cultivation,** 2005, p. 468.

GARCIA, R. A. F. **Use of ripeners to bring forward the maize harvest.** 22f. 2014. Dissertation (master's degree in agronomy). Paulista State University, Faculty of Agricultural and Veterinary Sciences. Jaboticabal, SP.

GILMAR, S. S. Sugar Sorghum: Agronomic and Industrial Technology for Food and Energy. **Agroenergia em Revista,** n. 3, p. 39. Aug. 2011
GOMES, A.; RODRIGUES, D.; OLIVEIRA, P. Sugar Sorghum: Agronomic and Industrial Technology for Food and Energy. **Agroenergia em Revista,** v. 2, n. 3, p. 26, Aug. 2011.

GURGEL, M. A. **Extended harvest using sugar sorghum. Dedini S/A Indústrias de Base,** 2010. Available at :
<http://www.dedini.com.br/index.php?option=com_docman&task=doc_view&gid=101&Iteid =40&lang=en>. Accessed on: 13 Aug. 2015

Introduction to plant biology. São Carlos: University of São Paulo, 2001. Available at: <http://biologia.ifsc.usp.br/bio3/outros/03-Fisiologia.pdf>. Accessed on: 24 August 2015.

KERBAUY, G.B. **Fisiologia Vegetal.** Rio de Janeiro: Guanabara Koogan, 2004.
LANDAU E. C.; GUIMARÃES D. P. **Embrapa Milho e Sorgo,** 2010. Available at: <http://www.cnpms.embrapa.br/publicacoes/sorgo_6_ed/zoneamento.htm> Accessed on: 12 Aug. 2015.

LAMAS, F. M. **Growth regulators.** In: EMPBRAPA AGROPECUÁRIA OESTE. Cotton:

Production technology. Embrapa Agropecuária Oeste, Embrapa Cotton, Dourados, 2001, p. 296.

LANDAU, E. C.; SCHAFFERT, R. E. **Zoning areas suitable for planting sugarcane sorghum during the sugarcane off-season in Brazil.** Agroenergia em Revista. Issue 3, p.20. 2011.

LARCHER, W. Plant ecophysiology. São Carlos: RIMA, 2000, p. 531.

LAVAGNINI, C. G.; DI CARNE, C. A. V.; CORREA, F.; HENRIQUE, F.; TOKUMO, L. E.; SILVA, M. H.; SANTOS, P. C. S. Plant Physiology - Gliberellin Hormone. **Electronic Scientific Journal of Agronomy.** Garça - São Paulo, v.25, n.1, p. 48-52, jun.2014.

LINDOMAR, R.; HAVANE, E. **Gibberellins.** S.d. Available at: < http://pt.slideshare.net/lindomarricardo7/seminario-de-fisiologia-vegetal>. Accessed on: 29 August 2015.

MACHADO, C. M. M. Ethanol production from sugar sorghum. **Agroenergia em Revista**, v. 2, n. 3, p. 27, Aug. 2011.

MACHADO, C. M. M. Sugar sorghum: development of industrial technology. **Agroenergia em Revista**, n. 3, p. 27-28. Aug. 2011.

MACHADO, J. R. J.; NAKAGAWA, C. A.; ROSOLEM, O. B. 1987. Sowing times for sugar sorghum in São Manuel and Botucatu, São Paulo State. Brazilian Agricultural Research.

MAGALHÃES, B. S. N.; VIERA, M. C. R. **Plant Hormones.** 2008. Available at: <http://www.projetofundao.ufrj.br/biologia/images/materiais/hormonios_vegetais_mariana_c abrera_barbara_neil.pdf>. Accessed on: 19 August 2015.

MAGALHÃES, P. C.; DURÃES, F.O.M.; SCHAFFERT, R. E. Physiology of the sorghum plant. Sete Lagoas, MG; Embrapa Milho e Sorgo. n.3, p. 46, 2000.

MANTOVANI, E. C. National Maize and Sorghum Research Centre. **Sorghum cultivation.** Production system 2. Sete Lagoas-MG, 2000.
MARCOCCIA, R. **The participation of Brazilian ethanol in a new perspective of the world energy matrix.** 2007. 95f. Dissertation presented to the postgraduate programme in Energy at the University of São Paulo. São Paulo, SP. 2007.

MAY, A. Good agricultural practices for growing sugar sorghum. **Agroenergia em Revista**, n. 3, p. 16-17, Aug. 2011.

MAY, A. **Embrapa Maize and Sorghum.** Growing sugarcane sorghum in sugarcane reform areas. Sete Lagoas, MG. 2013. Available at: <http://www.infoteca.cnptia.embrapa.br/infoteca/bitstream/doc/966886/1/circ186.pdf>. Accessed on: 17 August 2015.

MENDES, M. C. New technology for greater productivity and longevity in sugarcane plantations. 2010. Available at   : <www.ruralnoticias.com/FUTSite/default_processa.asp?elemento=noticia&id=1438> Accessed on: 25 September 2015.

MOLIN, R. Sowing row spacing in maize cultivation. **Castro, ABC Foundation for Agricultural Technical Assistance and Dissemination.** p.1 - 2, 2000.

OLIVEIRA, M. Sorghum is planted to produce ethanol in the sugarcane off-season. **Pesquisa FAPESP.** v. 193. p. 64-65, Mar. 2012.

OLIVEIRA, R. S. Jr.; CONSTATIN, J.; INOUE, M.H. **Biologia e Manejo de plantas.** Omnipax. Curitiba, PR. 2011. Available at : <http://omnipax.com.br/livros/2011/BMPD/BMPD-livro.pdf>. Accessed on: 13 Aug. 2015 PONTIN, J. C. Evaluation of plant ripeners in sugar cane. **Álcool e Açúcar,** São Paulo, n. 77, p. 16-18. 1995.

QUEVEDO, D.; PERREIRA, M. A.; CHAPIESKI, P. C.; SCOLARI, T.; BERTOLDI, W. P. **Ethylene.** 2011. Available at: <http://pt.scribd.com/doc/98589517/Etileno-e-Suas-Caracteristicas#scribd> Accessed on: 01 Sep. 2015.

RIBAS, P. M. **Sorghum:** Introduction and Economic Importance. Sete Lagoas, MG: EMBRAPA, 2003.

RODRIGUES, H. F. F. **Densities and Cutting Times of Sugar Sorghum Plants for Fodder and Ethanol Production.** 2010. 43f. Monograph presented to the agronomy course, Montes Claros State University. Janaúba, MG. 2010.

RODRIGUES, T. J. D.; LEITE, I. C. **Fisiologia vegetal:** Hormônio das plantas. Jaboticabal, SP: Funep, 2004.

SCHAFFERT, R. E.; GIACOMINI, F.; BORGONOVI, R. A. **Alcohol Production from Sugar Sorghum.** In: Seminário sobre energia de biomassa no nordeste, 1., 1978, Fortaleza. *Proceedings...* Fortaleza: UFC, 1978, p. 169-188.

SILVA, D. M; CARNEIRO, L. L.; MENDES, D. J.; SIBOV, S. T. Control of the auxins naphthalene acetic acid and indole butyric acid on the *in vitro* development of *Cyrtopodium saintlegerianum* Rchb. F. (ORCHIDACEAE). **Enciclopédia Biosfera,** Centro Científico Conhecer.

Goiânia, v. 9, n. 16; p. 52. 2013.

SILVA, E. A. A.; BRAGA, E. A.; BRITO, K. M.; V, F.; ANDRADE, A. C. Determination of abscisic acid levels during the ripening phase of *Coffea arabica* and *Coffea canephora* seeds. In: SIMPÓSIO DE PESQUISA DOS CAFÉS DO BRASIL, 4, 2005, Londrina, **Anais...** Brasília, D.F: Embrapa Café, 2005.

SILVA, T. T. A. **Quality of sorghum seeds (Sorghum bicolor L.) during ripening, drying and storage**. 125f. 2010. Thesis (PhD). Federal University of Lavras. Lavras, MG.

SORDI, R. A. Sugar sorghum for ethanol production: a view from the producer and the sugar cane mill. **Agroenergia em Revista**. n. 3, p. 31-32. Aug. 2011.

TEIXEIRA, C.G.; JARDINE, J.G. and BEISMAN, D. A. Utilisation of sugar sorghum as a complementary raw material to sugar cane for obtaining ethanol in micro-distilleries. Ciênc. Tecnol. Aliment. v.17, n.3. p. 248-251, 1997.

Buy your books fast and straightforward online - at one of world's fastest growing online book stores! Environmentally sound due to Print-on-Demand technologies.

Buy your books online at
www.morebooks.shop

Kaufen Sie Ihre Bücher schnell und unkompliziert online – auf einer der am schnellsten wachsenden Buchhandelsplattformen weltweit! Dank Print-On-Demand umwelt- und ressourcenschonend produziert.

Bücher schneller online kaufen
www.morebooks.shop

Printed by Books on Demand GmbH, Norderstedt / Germany